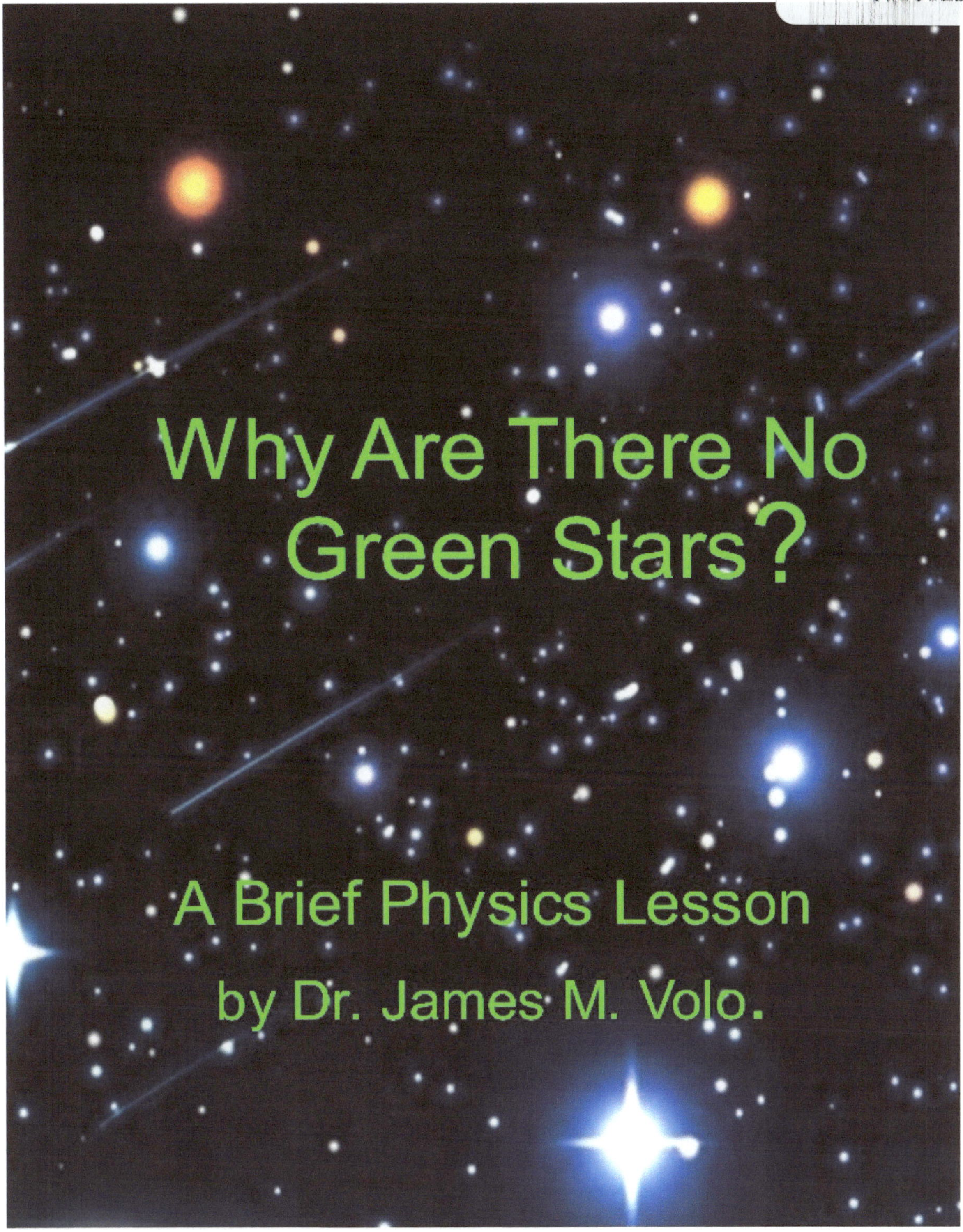

Why Are There No Green Stars?

A Brief Physics Lesson

by Dr. James M. Volo.

Black Body Radiation and the Perception of Color

The nighttime sky is filled with millions of bits of colored light. The average human can discern about 3000 individual stars in a clear sky, and with a simple telescope or binoculars they can "see" many times that number in white, blue-white, yellow, red, and variations of these colors. Without special filters none of these bits seem to be green, yet the greens are there — largely hidden from human view. This brief monograph will attempt to explain why there are no green stars.

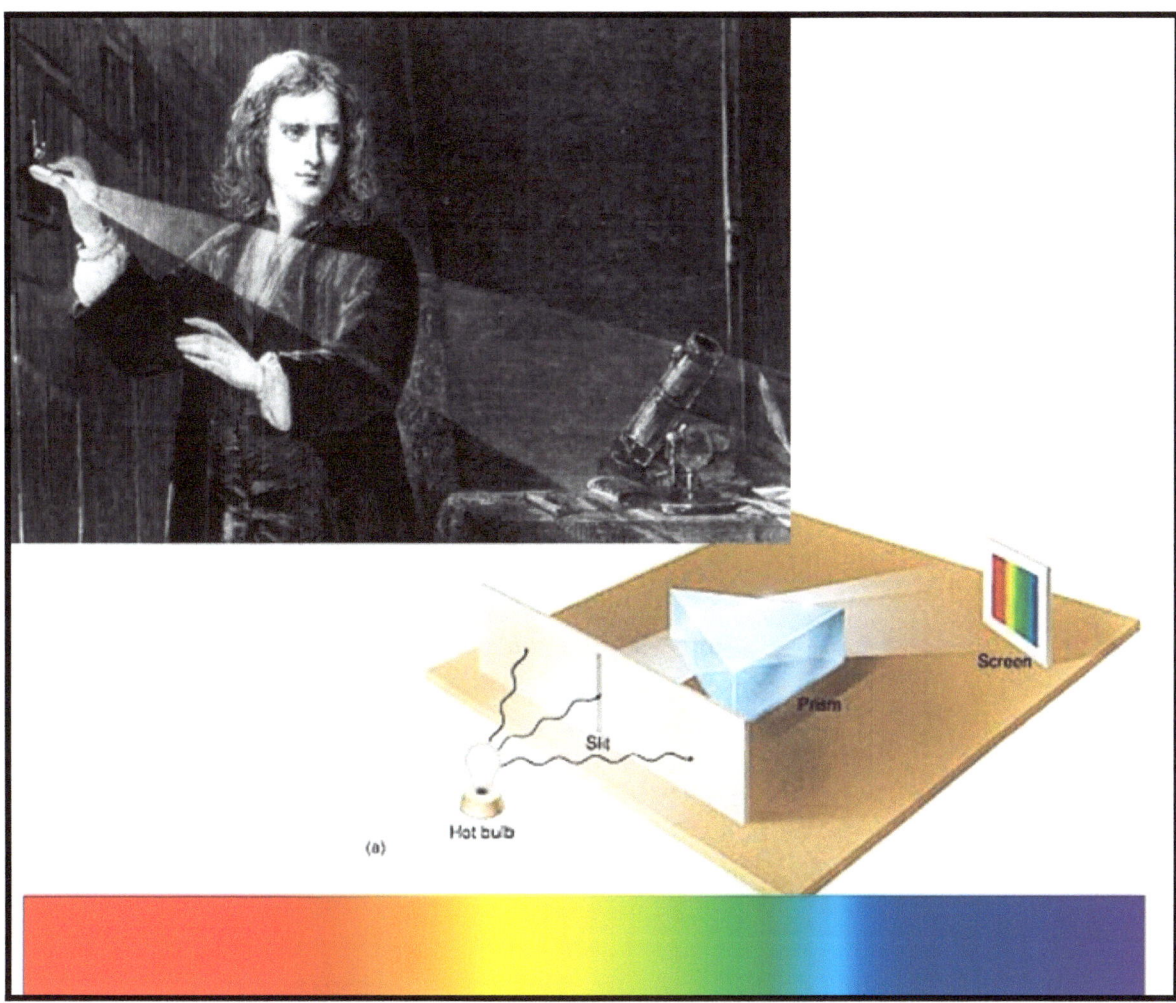

Sir Isaac Newton famously demonstrated that sunlight (full spectrum) could be broken down into its constituent colors when passed through a dispersive glass prism. He termed the parts of the visible spectrum red, orange, yellow, green, blue, indigo and violet bands. The Sun is correctly described as a medium-sized yellow star, but all these colors are commonly seen in a rainbow made from sunlight.

It is known today the so-called white light has the same parts, but these have been broken down into hundreds, thousands, and millions of "colors" distinguished by their wavelengths — reds being long decreasing to shorter blues. There are many more reds than any other band or sub-band. Just check the color settings on your computer; some have 2 million color settings. Greens fall into one of the narrowest of these bands. If green shows up in the spectrum of the sun, and the sun is a star, why don't we see it? We notice its yellowness but never its greenness. As Shakespeare wrote: "The fault … is not in our stars, it is in ourselves."

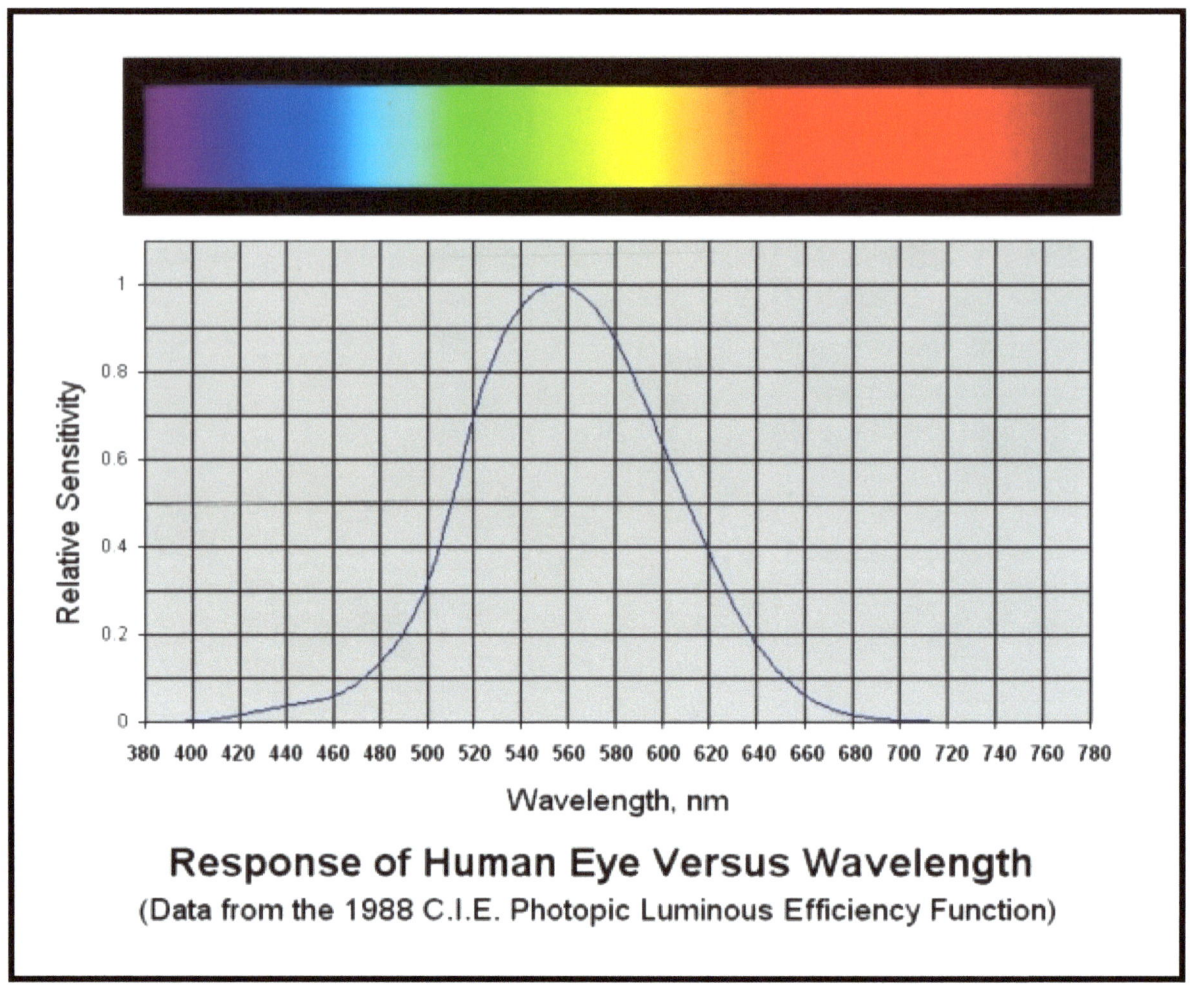

Response of Human Eye Versus Wavelength
(Data from the 1988 C.I.E. Photopic Luminous Efficiency Function)

Ironically, green is one of the visible colors to which humans respond the best. Along with yellow, the human response in terms of sensitivity to green is better than that to the blue or red ends of the visible spectrum. In the past, for instance, fire engines were painted red, but today they are increasingly done in yellow-green (lime). Over a century ago, red was a nice bright color that set fire apparatus apart from most vehicles on the road, but these private vehicles were mostly black. Part of the response to red was psychological rather than physiological. Fluorescent orange was chosen for hunters' vest for its obvious contrast with green foliage, yet it is less effective against the reds and oranges of autumn.

A 2009 study by the U.S. Fire Administration (USFA), a division of the Federal Emergency Management Agency (FEMA), concluded that fluorescent colors, including yellow-green and orange, are easiest to spot in daylight and under dim conditions, making lime shades easiest to see in low lighting. However, if people in a particular community don't associate the color lime with fire trucks, then yellow-green emergency vehicles may not actually be as conspicuous as intended.

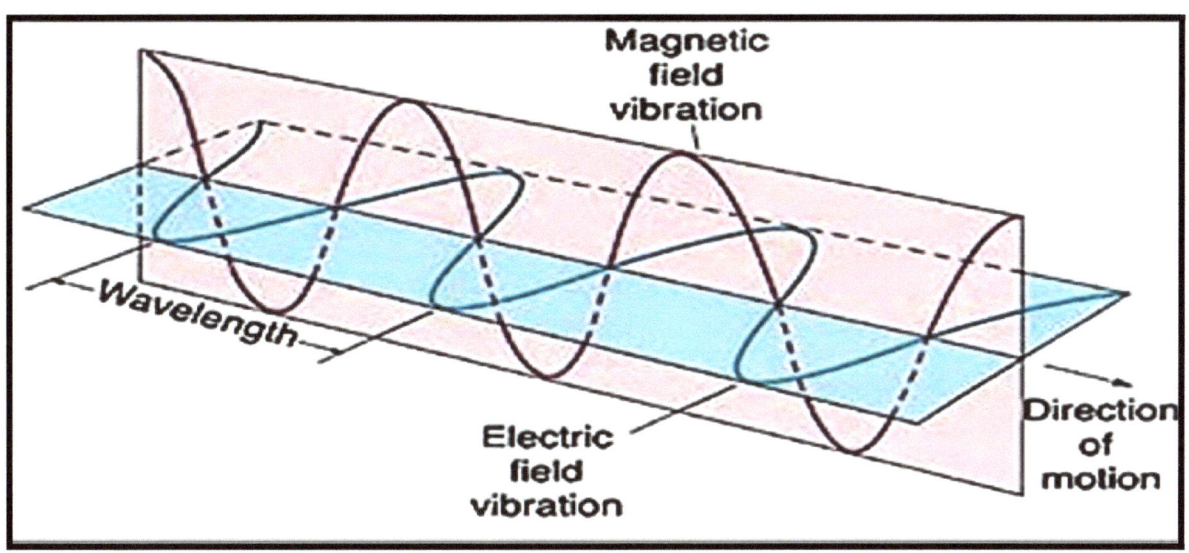

All forms of visible light are electromagnetic in nature regardless of color. EM waves have equally strong electric and magnetic fields traveling at right angles to one another and to their direction of propagation (motion). These fields stimulate a response in the human eye that is perceived as light-dark, contrast, outline, form, and color. Yet the span of EM waves in the universe is much greater than the narrow confines of VISIBLE light.

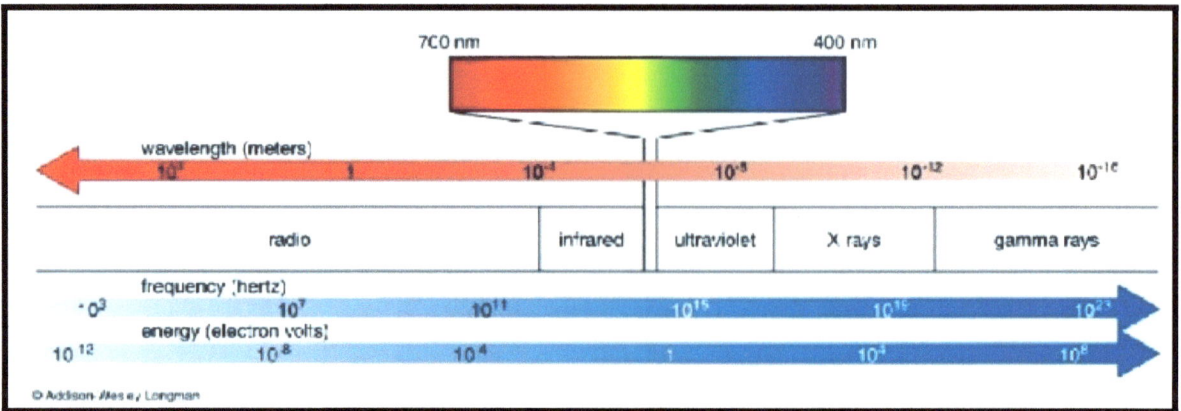

Humans do not easily perceive most of the Electromagnetic spectrum. Some wavelengths of Infrared are felt as heat. Rattlesnakes can "see" in parts of the infrared and many insects "see" in the ultraviolet. Wavelengths in this diagram are measured in nanometers (nm) and their associated frequencies in Hertz (Hz). Generally, the long reds have less energy than the short blues all other things being equal (ROYGBIV). This is true for the entire EM spectrum. On this diagram, things to the left have less energy than those to their right. The diagram is often found in reverse.

Many of the characteristics of light can be compared to the action of waves in water. Waves from a central source tend to spread out in all directions equally, and they maintain their wavelength, frequency, and speed over distance. Visible light and all EM waves travel at 186,000 miles per second (3×10^8 m/s) in free space. Light waves generally slow down in optically dense transparent materials like water, glass, and parts of the eye. The transparent cornea and lens of the human eye serve to form images on the retina where bio-electrical pulses are sent to the visual cortex of the brain for interpretation.

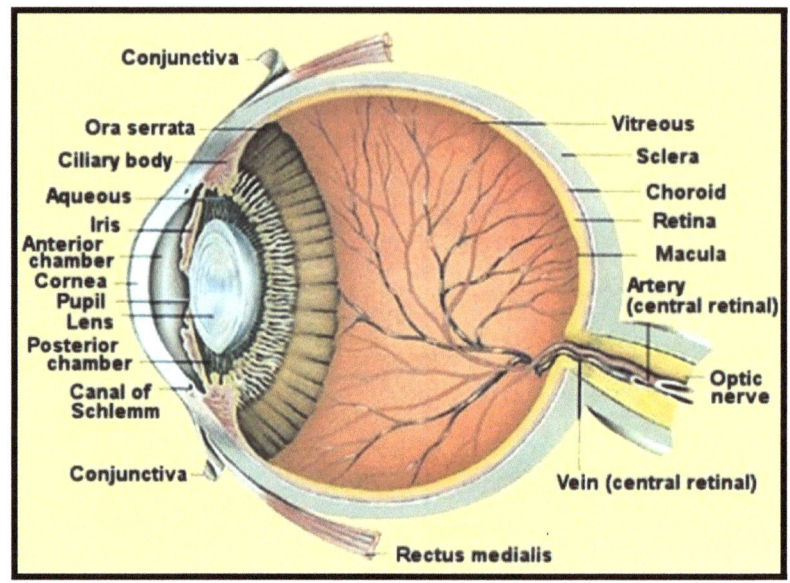

Contrary to what might be thought, the human eye is not a perfect sphere, but is made up of two differently shaped pieces, the cornea and the sclera. A ring called the limbus connects these two parts. The part of the eye that is seen is the iris, which is the colorful part of the eye. In the middle of the iris is the opening called the pupil, the black "dot" that apparently changes size when exposed to different intensities of light. The cornea covers these elements, but is transparent. The fundus is on the opposite of the pupil, but inside the eye and cannot be seen without special instruments. The optic nerve is what conveys the signals of the eye to the brain.

Whenever the eye moves, even just a little, it automatically readjusts the exposure by adjusting the iris, which regulates the size of the pupil. This is what helps the eye adjust to dark places or really bright lights. The lens of the eye is similar to one in glasses or cameras. It is made of clear cartilage. The human pupil and iris act just like a camera. The pupil serves this function, and the iris is the aperture stop. The different parts of the eye have different refractive indexes, and this is what bends the rays to form an image. The cornea provides two-thirds of the power to the eye. The lens provides the remaining focusing power. The image passes through several layers of the eye, but happens in a way very similar to that of a convex lens. When the image finally reaches the retina, it is inverted, but the brain corrects this inversion. The human eye is more sensitive to intensity changes than to color changes, which is why it is acceptable to use black and white photography in place of color, and why people can still distinguish everything in the photo without colors.

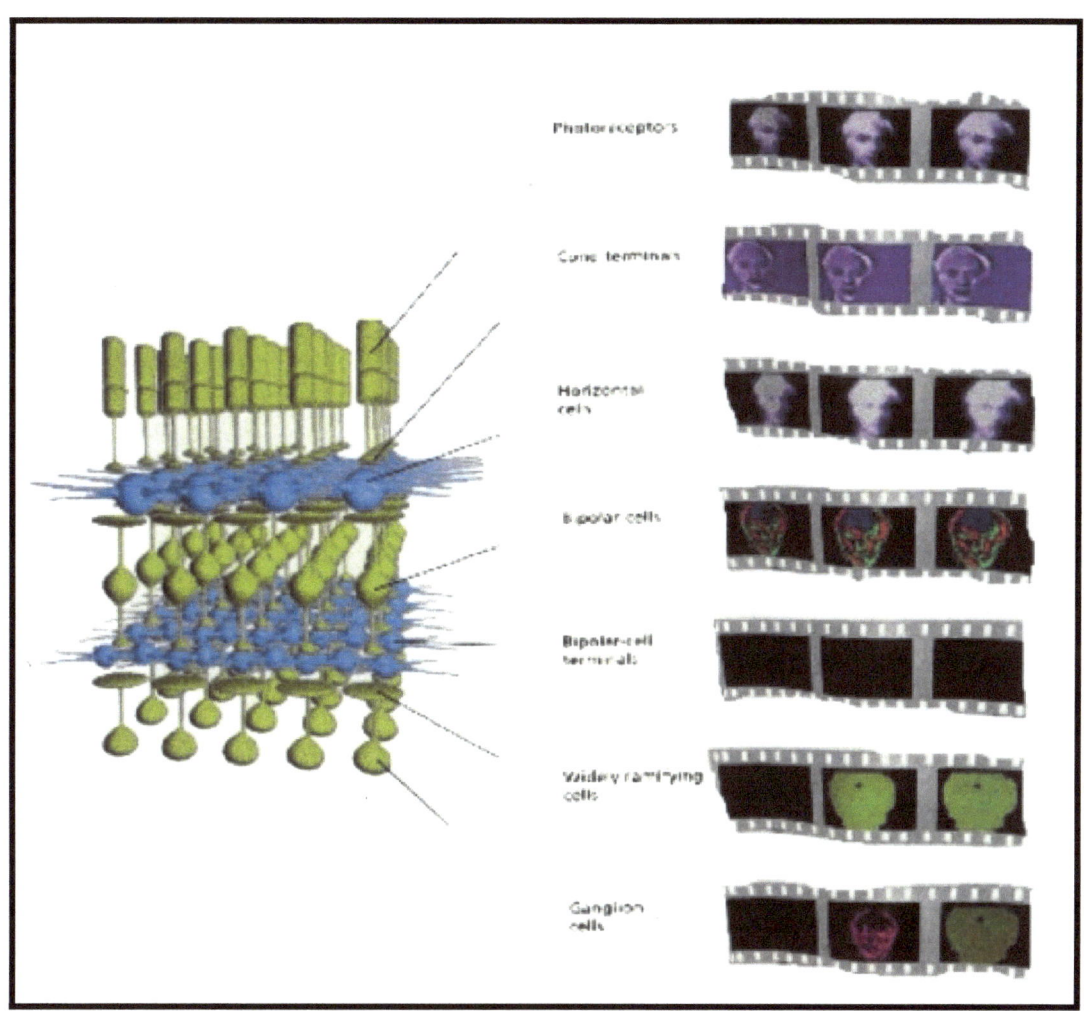

The eye and brain combine different portions of an image such as form, color, motion, and shading into a complete image in much the same way that a series of photographs can be combined to create a finished picture. Scientists have known for more than 200 years that vision begins with a series of chemical reactions when light strikes the retina, but the specific chemical processes have largely been a mystery. At the center of the theory is the signaling of rhodopsin to transducin. Rhodopsin is a pigment in the eye that helps detect light. Transducin is a protein (sometimes called "GPCR") that ultimately signals the brain that light is present.

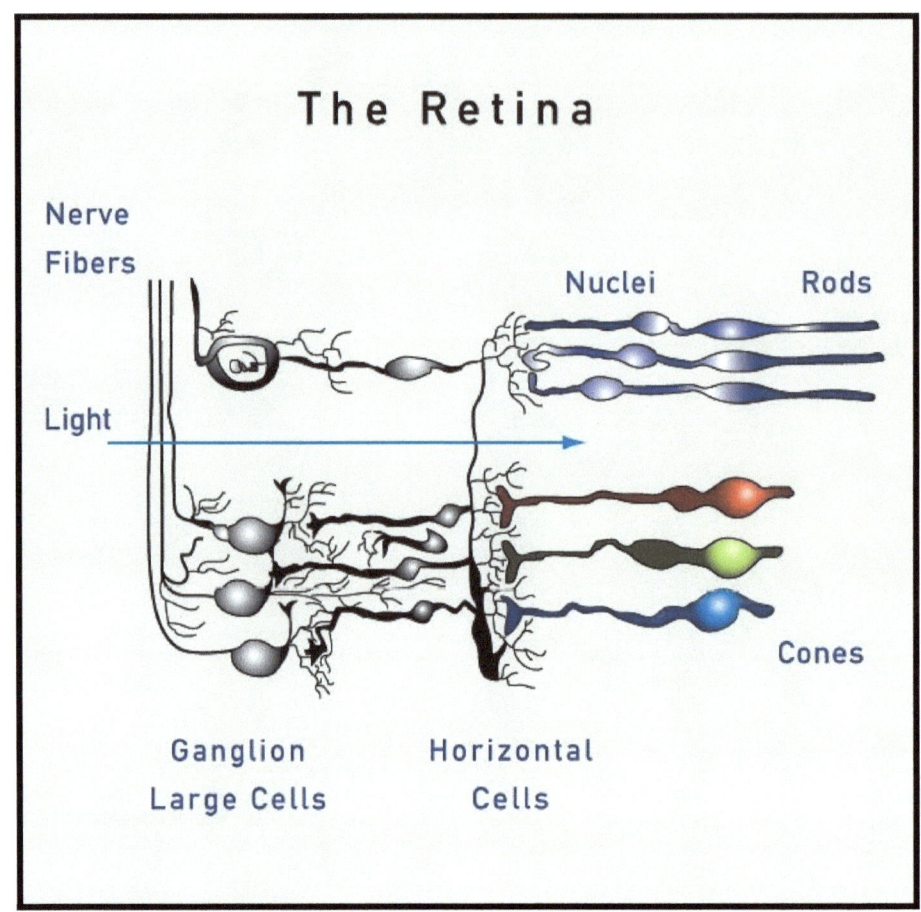

The retina consists of a very thin layer of nerve cells and contains two types of photoreceptors, rods and cones. The numbers of rods and cones vary over the surface of the retina. Rods are located mainly in the peripheral retina and are absent in the middle of the fovea (the center of the retina). Cones are located throughout the retina but are concentrated in the very center of the retina. The rods, which are more numerous than cones, are responsible for our vision in dim light but don't function well in bright light. Cones (specifically sensitive to red, blue, or green) are active at high light levels and allow us to see color and fine detail. Rods account for our night vision but cannot distinguish color. Ganglions are also located in the retina.

Ganglions are a type of neuron located in the retina that receives signals via various intermediate cells from the cones and rods. They are the cells that transmit the information to the brain. The visual cortex of the brain is a part of the cerebral cortex that processes visual information. The primary visual cortex is the most studied visual area in the brain. It is located in the occipital lobe in the back of the head. Both hemispheres of the brain contain a visual cortex; the visual cortex in the left hemisphere receives signals from the right visual field, and the visual cortex in the right hemisphere receives signals from the left visual field. Those areas identified as having a role in color vision processing are collectively labeled visual area 4 (V4). The exact mechanisms, location, and function of V4 are still being investigated.

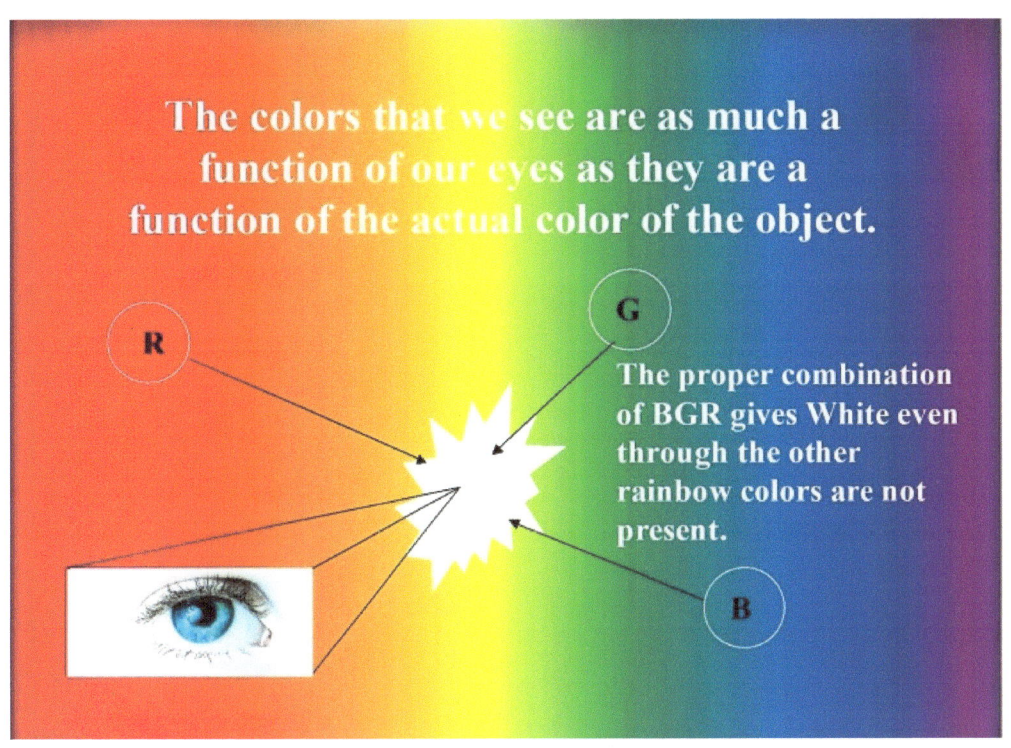

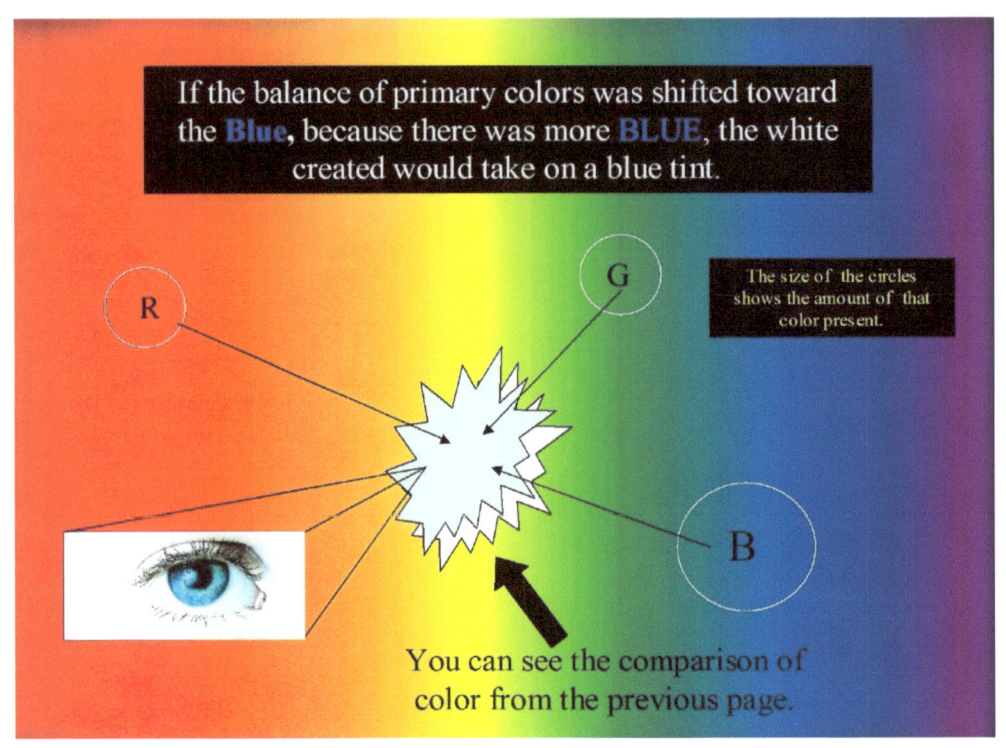

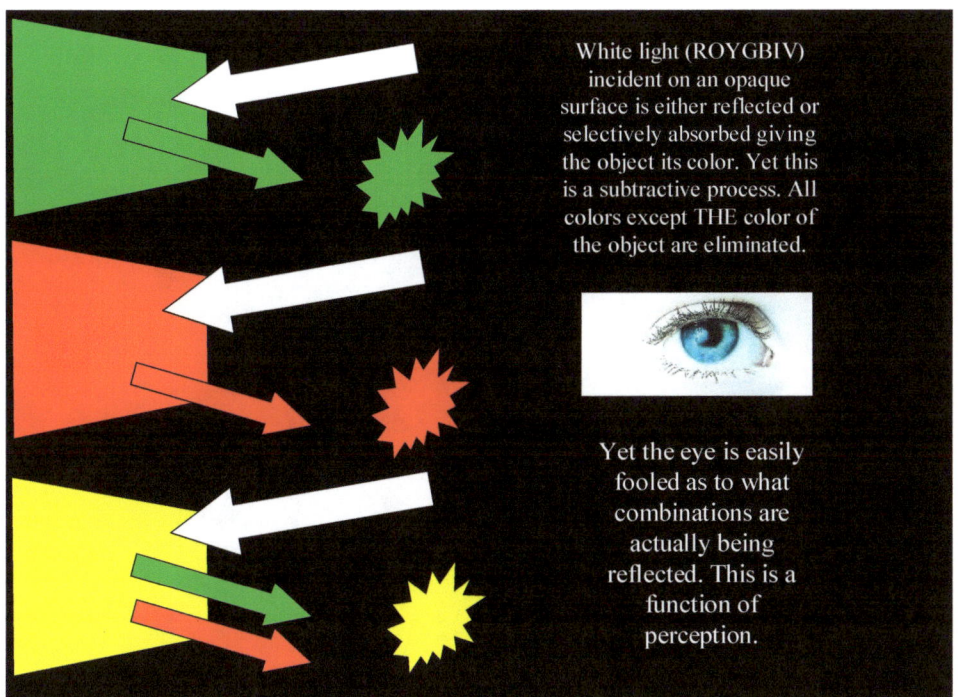

A subtractive color model explains the mixing of a limited set of dyes, inks, paint pigments or natural colorants to create a wider range of colors, each the result of partially or completely subtracting (that is, absorbing) some wavelengths of light and not others. The color that a surface displays depends on which parts of the visible spectrum are not absorbed and therefore remain visible. If the incident light is other than white, our visual mechanisms are able to compensate well, but not perfectly, often giving a flawed impression of the "true" color of the surface.

RYB (Red, Yellow, Blue) is the formerly standard set of subtractive primary colors used for mixing pigments. It is used in art and art education, particularly in painting. It predated modern scientific color theory. In late 19th and early to mid-20th century commercial printing, use of the traditional RYB terminology persisted even though the more versatile CMY (Cyan, Magenta, Yellow) triad had been adopted, with the cyan sometimes referred to as "process blue" and the magenta as "process red."

In modern color printing, the usual primary colors are cyan, magenta and yellow (CMY). Cyan is the complement of red, meaning that the cyan serves as a filter that absorbs red. The amount of cyan applied to a white sheet of paper controls how much of the red in white light will be reflected back from the paper. Ideally, the cyan is completely transparent to green and blue light and has no effect on those parts of the spectrum. Magenta is the complement of green, and yellow the complement of blue. Combinations of different amounts of the three can produce a wide range of colors with good saturation.

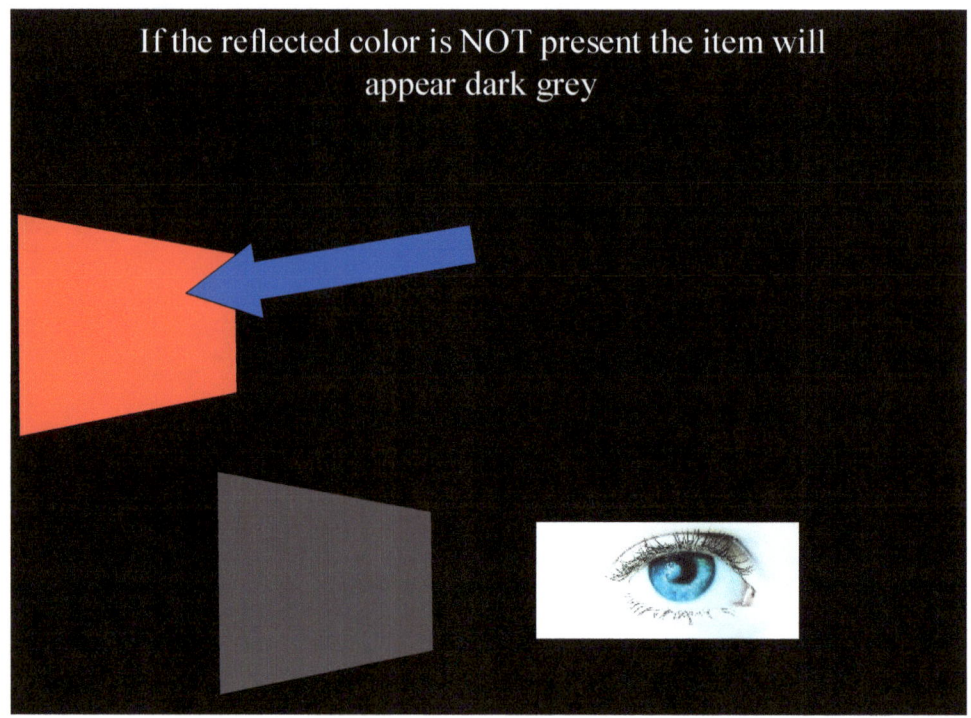

What color would a red apple appear in blue light? All plants and vegetables absorb visible light and reflect the colors that we see, so green trees reflect green light, yellow meadows absorb all colors except yellow. A red apple would absorb most of any blue light and appear grey. In addition, a green leaf appears dark grey at night. Each of these examples supports the subtractive color theory of reflection.

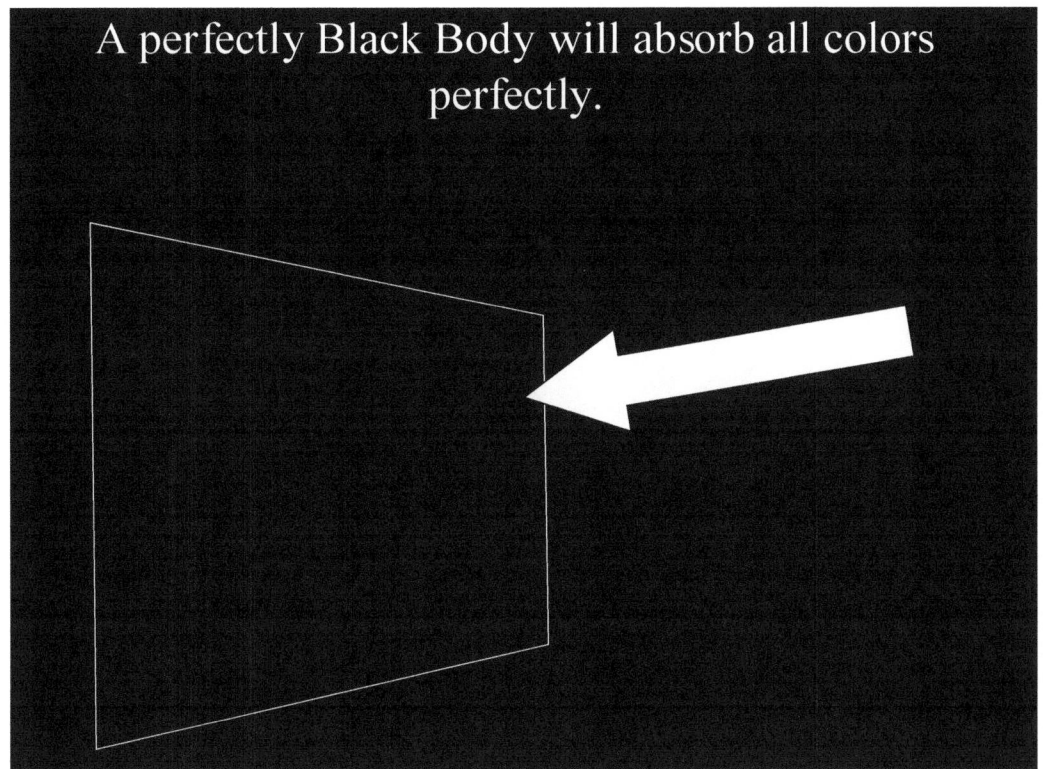

Does the energy absorbed by a black body dissipate, heating up, or is it instead converted into some other form? A black body is an ideal body, which allows the whole of the incident radiation to pass into itself (without reflecting the energy) and absorbs within itself this whole incident radiation (without passing on the energy). This property is valid for radiation corresponding to all wavelengths and to all angles of incidence. Therefore, the black body is an ideal absorber of incident radiation. To stay in thermal equilibrium, it must emit radiation at the same rate as it absorbs it so a black body also radiates well. (Stoves are black.)

Somewhat counterintuitively, black bodies are also considered perfect radiators and don't need to appear black. As the black body absorbs energy its temperature increases. Heat an object up to about 1500 degrees and a dull red glow appears. The object is red hot. Heat something up to about 5000 degrees, near the temperature of the sun's surface, and it radiates well throughout the visible spectrum. It is white hot. At a particular temperature the black body begins to emit the maximum amount of energy possible for that temperature.

It can be said that the stars radiate like blackbody radiators. This is important because it means that the theory for blackbody radiators can be used to infer things about stars. Blackbody radiation does not depend on the type of object emitting it. The entire spectrum of blackbody radiation depends on only one parameter, the temperature.

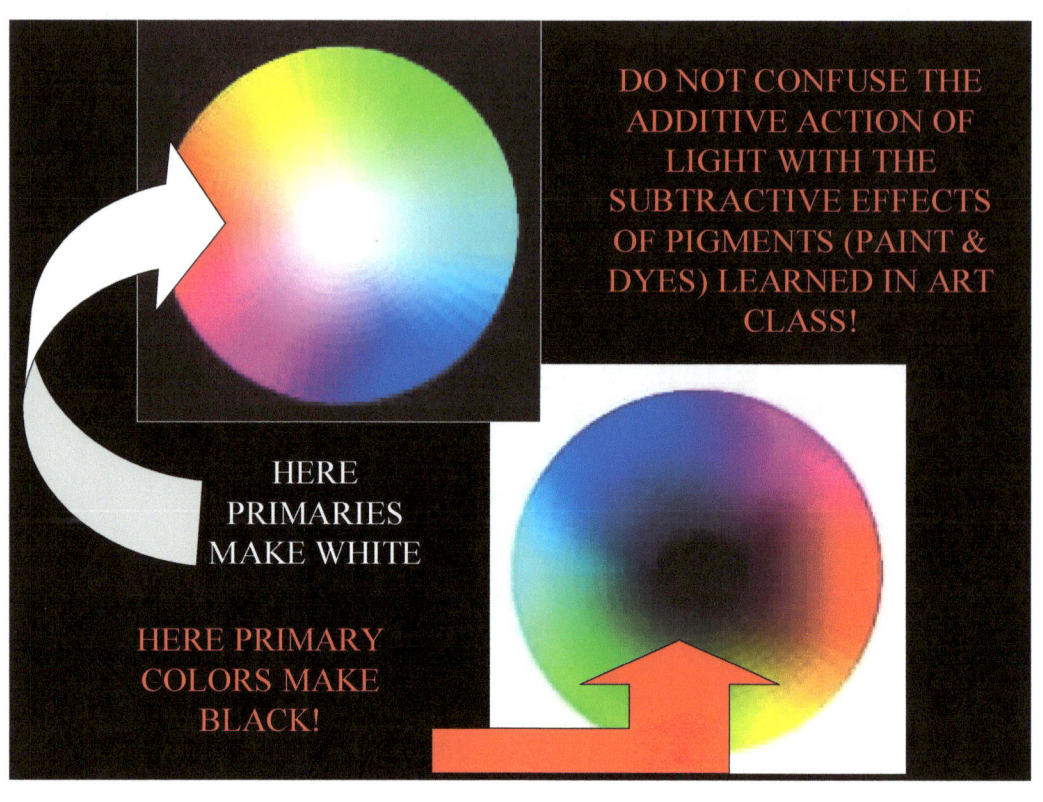

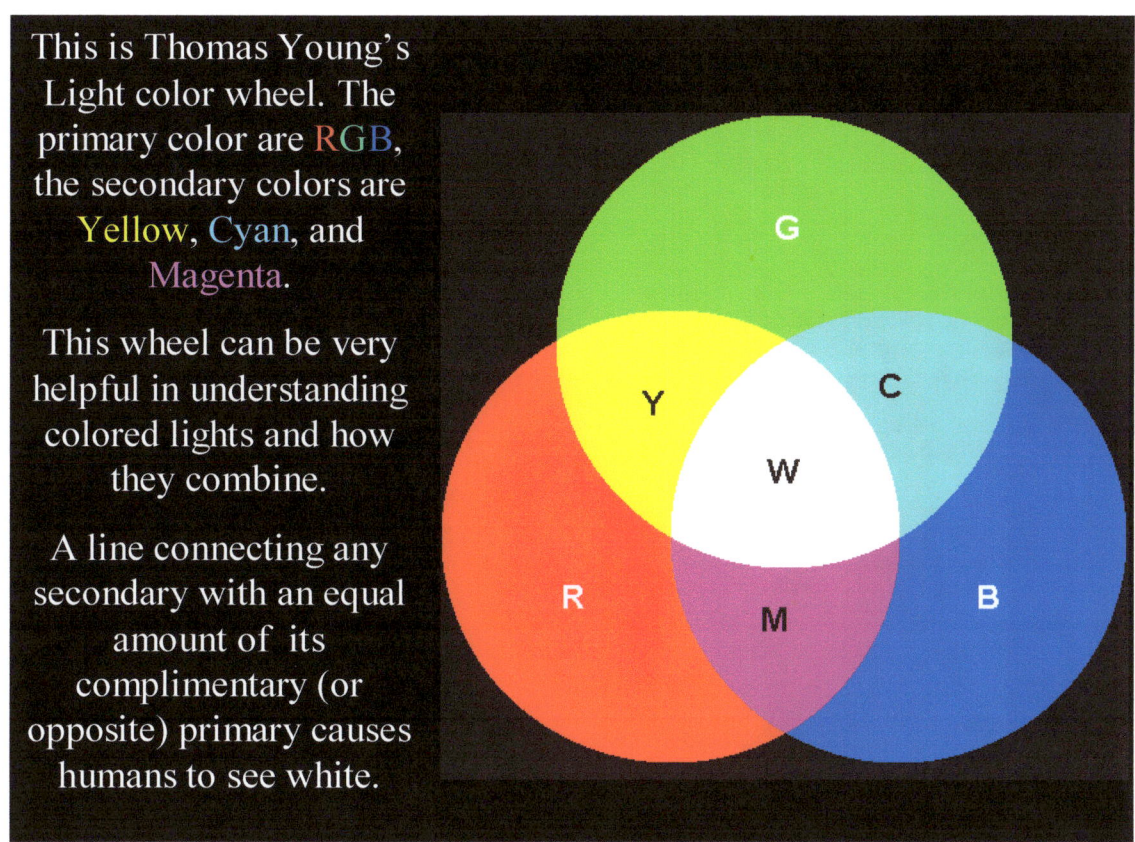

This is Thomas Young's Light color wheel. The primary color are RGB, the secondary colors are Yellow, Cyan, and Magenta.

This wheel can be very helpful in understanding colored lights and how they combine.

A line connecting any secondary with an equal amount of its complimentary (or opposite) primary causes humans to see white.

According to the trichromatic theory of color vision, also known as the Young-Helmholtz theory of color vision, there are three receptors in the retina that are responsible for the perception of color. One receptor is sensitive to the color green, another to the color blue and a third to the color red. These three colors can then be combined to form any visible color in the spectrum. Both Thomas Young and Hermann von Helmholtz contributed to the trichromatic theory of color vision.

The theory began when Young proposed that color vision results from the actions of three different receptors. As early as 1802, Young suggested that the eye contained different photoreceptor cells that were sensitive to different wavelengths of light in the visible spectrum. Young wrote: "As it is almost impossible to conceive each sensitive point of the retina to contain an infinite number of particles, each capable of vibrating in perfect unison with every possible undulation [color], it becomes necessary to suppose the number limited, for instance to the three principal colors."

It was later in the mid-1800s that Helmholtz expanded upon Young's original theory and suggested that the cone receptors of the eye were either short-wavelength (blue), medium-wavelength (green), or long-wavelength (red). The definite identification of the three receptors responsible for color vision did not occur until more than 70 years after the proposal of the theory of trichromatic vision. Note how red and green (both dark colors) make a vibrant yellow. A lack of sensitivity in one or more types of cone is a common cause of color blindness.

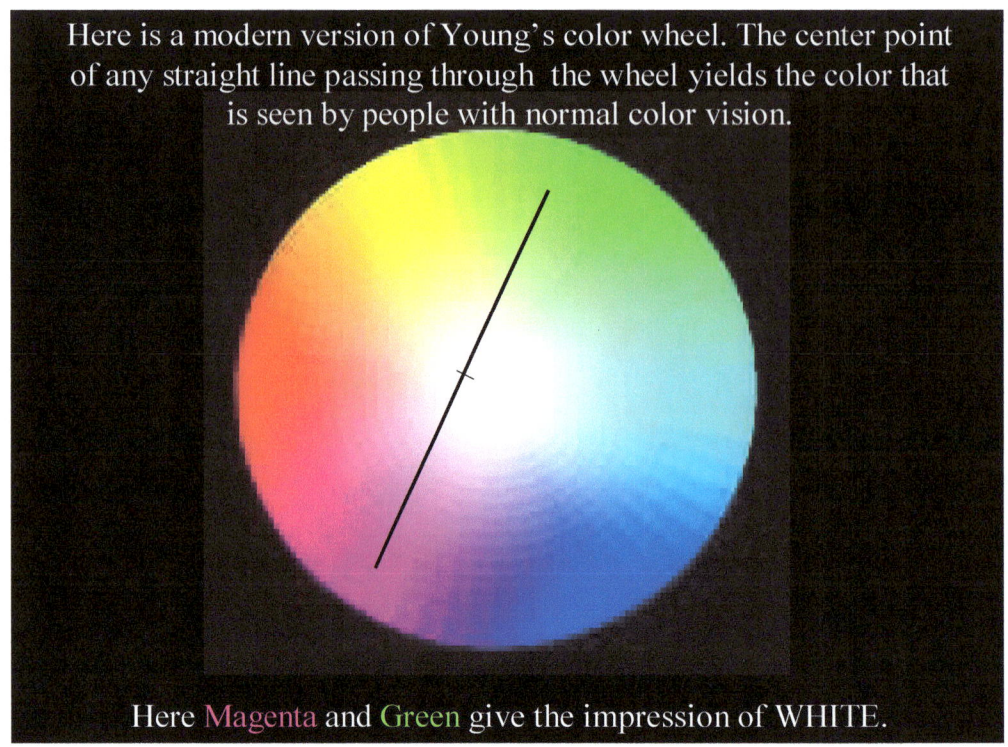

Any two areas of this color wheel can be selected with the ends of the line on the component color and the center of the line giving the perceived color.

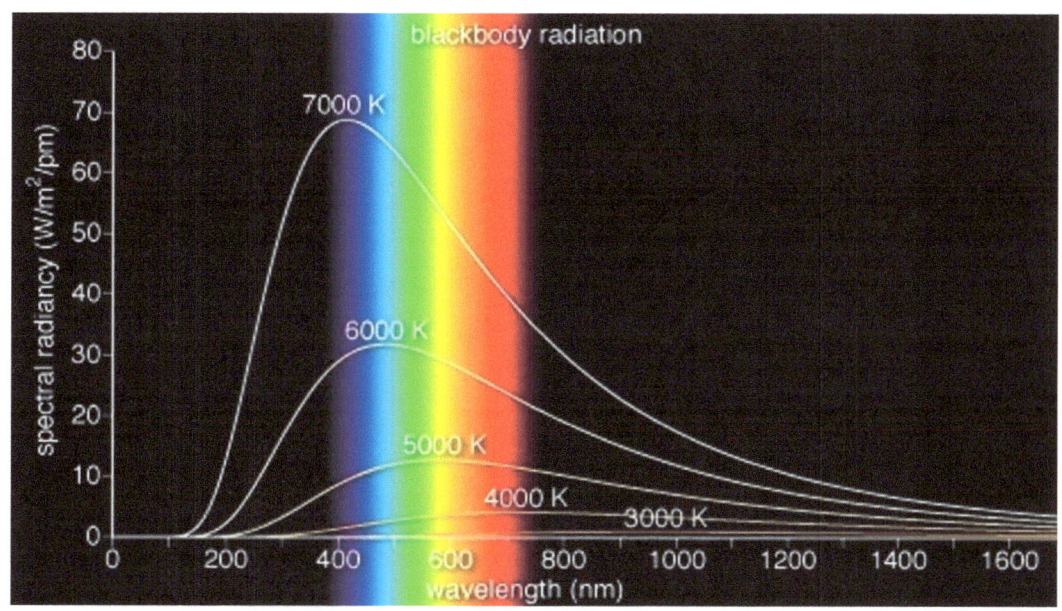

Stars at different temperature all radiate the full color spectrum, but the amount of each color present changes. Generally, hotter stars put out more color and shorter wavelengths. The red, green and blue in each case creates white, but in cooler stars the red dominates, in hotter stars red and green combine to form a plausible yellow, and in the hottest stars, blue tints the RGB white. There is never enough green to tint a star that color. Most of it combines with red as a perception in our eyes to form additional yellow. This is essentially why there are no stars tinted green, but there are many stars tinted blue, yellow, and red.

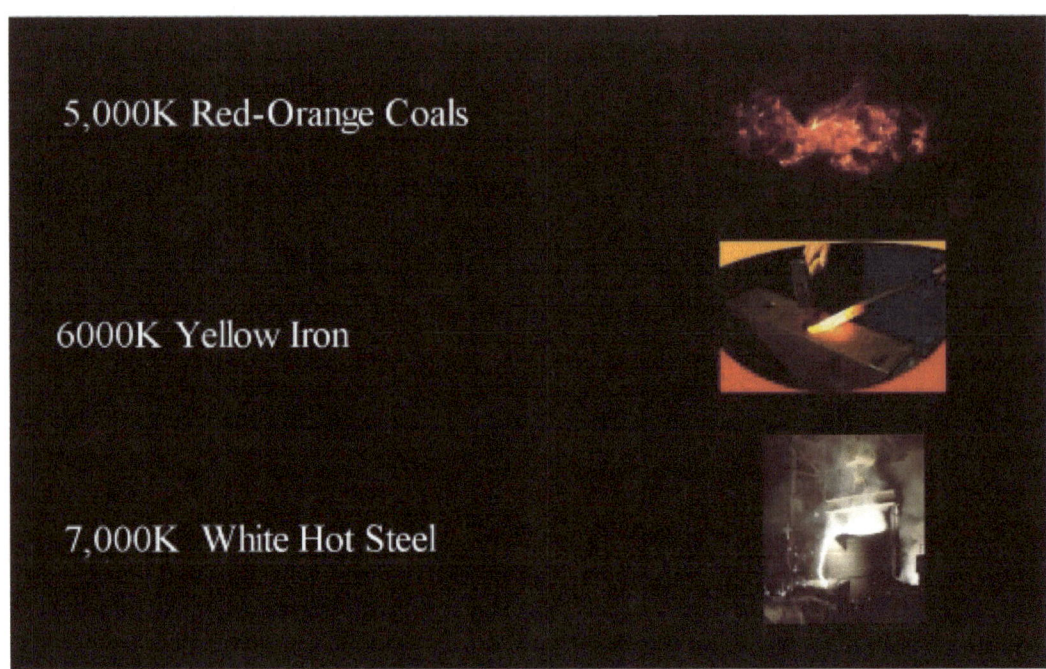

Black Body Radiation is the thermally induced electromagnetic radiation within or surrounding a body in thermodynamic equilibrium with its environment, or emitted by a black body (an opaque and non-reflective body). It has a specific spectrum and intensity that depends only on the body's surface temperature. These tend to run from red to orange to yellow to white to blue-white. No greens! Our Sun has a surface temperature that makes it appear yellow, but its internal temperature is many millions of degrees.

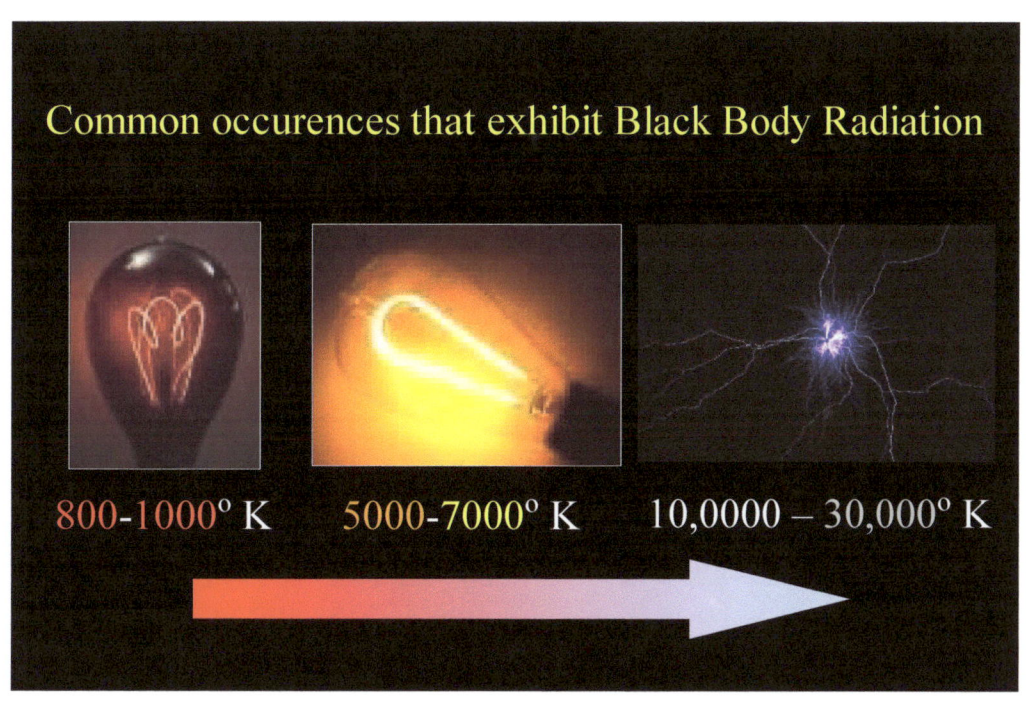

Here we can see how the color of a star can be used to determine temperature.

The star colors go from:
O – Blue
B – Blue White
A - White
F – Yellow White
G – (Our Sun) Yellow
K – Orange Red
M - Red

SPECTRAL CLASS	SURFACE TEMPERATURE (K)	PROMINENT ABSORPTION LINES	FAMILIAR EXAMPLES
O	30,000	Ionized helium strong, multiply ionized heavy elements, hydrogen faint	
B	20,000	Neutral helium moderate, singly ionized heavy elements, hydrogen moderate	Rigel (B8)
A	10,000	Neutral helium very faint; singly ionized heavy elements; hydrogen strong	Vega (A0), Sirius (A1)
F	7,000	Singly ionized heavy elements; neutral metals; hydrogen moderate	Canopus (F0)
G	6,000	Singly ionized heavy elements, neutral metals; hydrogen relatively faint	Sun (G2), Alpha Centauri (G2)
K	4,000	Singly ionized heavy elements, neutral metals strong, hydrogen faint	Arcturus (K2), Aldebaran (K5)
M	3,000	Neutral atoms strong, molecules moderate, hydrogen very faint	Betelgeuse (M2), Barnard's Star (M5)

Apparent Magnitude

(useful for describing how bright objects appear from the Earth)

The original magnitude system of Hipparchus had:

magnitude 1 – the brightest stars
magnitude 2 …
magnitude 3 …
magnitude 4 …
magnitude 5 …
magnitude 6 – the faintest stars

Today the magnitude system has been extended to include much fainter and brighter objects.

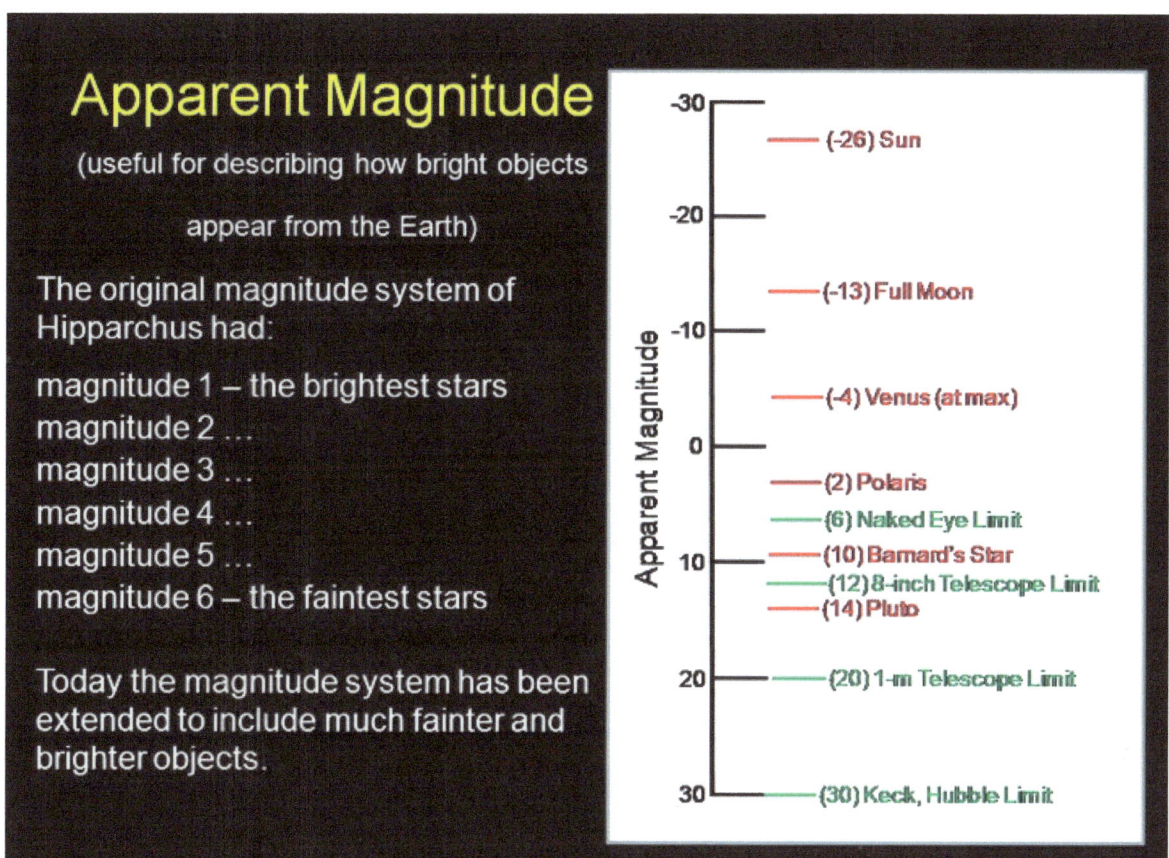

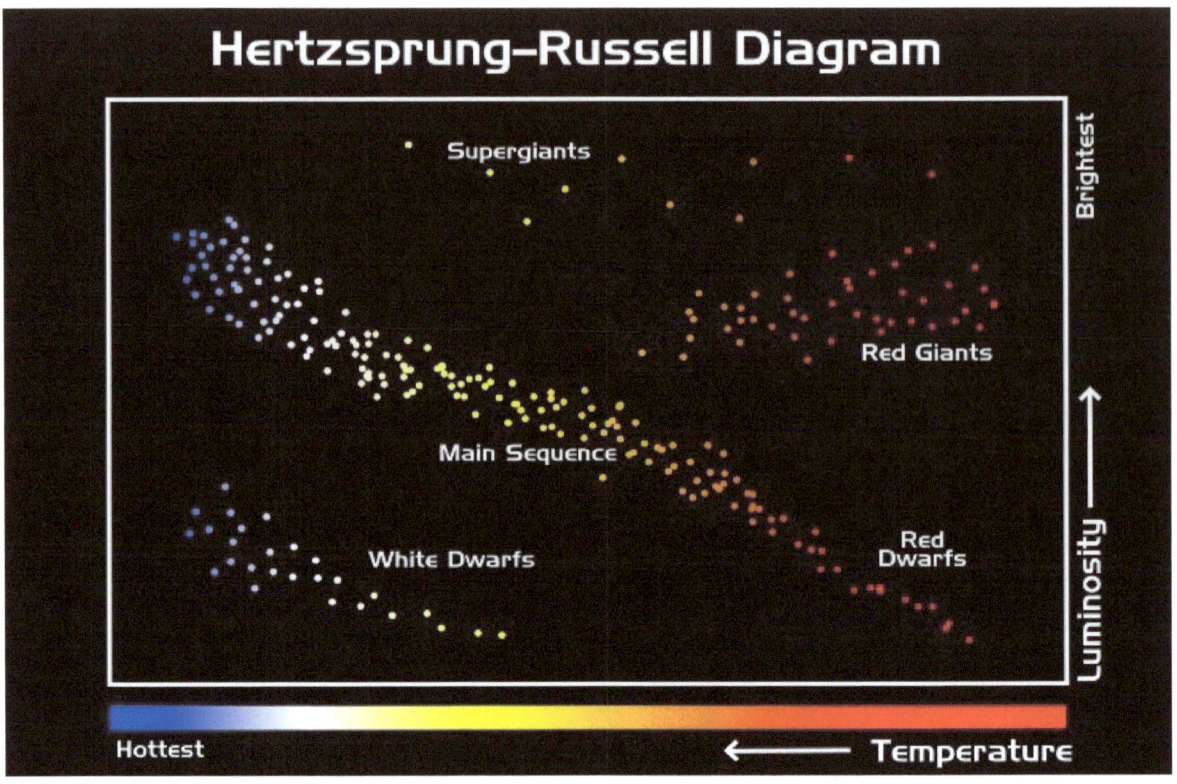

The Hertzsprung–Russell diagram, abbreviated H–R diagram, is a scatter plot of stars showing the relationship between their absolute magnitudes or luminosities versus their stellar classifications or effective surface temperatures. More simply, it plots each star on a graph measuring the star's brightness against its temperature (color). It does not map any locations of stars.

The diagram was created circa 1910. There are several forms of the Hertzsprung–Russell diagram, and the nomenclature is not very well defined. All forms share the same general layout: stars of greater luminosity are toward the top of the diagram, and stars with higher surface temperature are toward the left side of the diagram. Our Sun sits quite in the middle of the Main Sequence.

Contemplation of the diagram initially led astronomers to speculate that it might demonstrate stellar evolution, the main suggestion being that stars collapsed from red giants to dwarf stars, then moving down along the line of the main sequence in the course of their lifetimes. Stars were thought therefore to radiate energy by converting gravitational energy into radiation.

www.ingramcontent.com/pod-product-compliance
Lightning Source LLC
Chambersburg PA
CBHW051838210526
45473CB00005B/1937